MOUTHS AND NOSES

WRITTEN BY SALLY MARKHAM-DAVID
ILLUSTRATED BY TREVOR RUTH

Contents

Introduction	4
Finding Food	6
Catching Prey	10
Feeding	12
Fishing	14
Digging	16
Special Uses	18
Index	22

Introduction

When we think about the uses of mouths and noses, we probably think first of smelling and eating, breathing and talking. But animals also use their mouths and noses for many other purposes such as grooming, climbing, digging, and communicating. Many animals use their mouths for carrying their young.

Some animals, like the anteater, depend on their noses for finding food. The elephant can even use its nose for taking a bath! There are many different ways in which mouths and noses have been adapted to suit each animal's way of life.

The beaks of birds may be long and thin, or short and broad, depending on the food they eat and where they live. Beaks may be used for stabbing or tearing, pecking or probing. They are used as a tool for nest building and for the important job of keeping feathers in good condition.

Inside the mouths there are different kinds of tongues. They may be coiled like a butterfly's, rough like a cat's, or covered with teeth like a snail's.

Finding Food

To some animals their sense of smell is very important for finding food, while other animals use that sense very little.

The long nose of the giant anteater has a sense of smell which is forty times better than a human nose. The anteater has small eyes and ears so it relies on its sense of smell to find its food. With its sticky tongue as long as your arm, the anteater can sweep up thousands of ants at a single meal.

The tapir, a relative of the rhinoceros, has a short trunk for a nose. The tapir can move its trunk over the ground in search of food without moving its head. The tapir can draw its trunk in and out like an accordion. It pulls the trunk in while feeding, to keep the trunk out of the way.

Most birds have a poor sense of smell, but the kiwi, which is found only in New Zealand, finds its food by scent. The kiwi cannot fly, and it feeds by probing into soft soil with its bill. It has nostrils at the tip of its long bill, and these are used to help the kiwi find worms and insects in the soil.

Moles live mostly underground so their eyes are of little use. But the mole has very strong senses of smell and touch, and it feels with hairs on the tip of its snout. The star-nosed mole has a ring of fleshy tentacles on its snout and the mole waves these about in search of food.

Catching Prey

Many animals use their mouths as weapons for catching and killing their prey.

Dogs and cats, including wolves and leopards, have very strong canine teeth for stabbing their prey. Then they shred the flesh with their molars, or back teeth. The cats also have rough tongues that help to scrape the flesh off the bones.

Frogs catch insects by flicking their tongues in and out at great speed. The tongue is attached to the front of the mouth and folds backward. A frog's eyeballs bulge down into the roof of its mouth and when a frog swallows, it uses its eyeballs to push food down its throat. This is why a frog blinks every time it swallows.

Archerfish spurt jets of water from their mouths to shoot down insects that are crawling on leaves and stems hanging over the water. They can hit a target up to six feet above the surface.

The sloth bear uses its lips like a vacuum cleaner. When the bear has broken into a termites' nest with its claws, it puts its head into the nest, and forms its long, loose lips into a tube to suck up the termites. The bear closes its nostrils to keep termites out of its nose.

Feeding

The blue whale, the largest animal in the world, is toothless. Instead of teeth, it has two sets of hairy plates, called baleen. These hang from the roof of the whale's mouth, and work like sieves. The blue whale's main food is krill, a tiny shrimplike animal. As the whale takes water into its mouth and pushes it out again, krill get caught in the baleen.

Flamingos wade with their heads down and their bills upside down in the water. Their tongues draw water in and push it out again through a bristly comb along the edges of their bills. This traps small prey.

Fishing

Many animals have excellent fishing gear. Just as we all have our different ways of catching fish, birds have many different methods, too.

The pelican catches fish by scooping them up into the baggy pouch which is part of its beak. Skimmers scoop up fish, too, but they do it while flying. They skim along just above the sea, with their lower beaks cutting through the water. Gannets dive to catch fish, and their nostrils have coverings to keep the water out when they plunge into the sea.

Cormorants swim after fish, grabbing them from behind. A hook at the tip of the cormorant's beak helps it to hold onto the slippery fish. Herons and kingfishers often spear their prey, and a heron's bill has sawlike edges to prevent fish escaping. Penguins hold onto their catch with their rough tongues.

The alligator turtle has a forked tongue with two fat tips that look like worms. The turtle wriggles these to attract small fish, which it then snaps up.

Digging

Many animals dig with their limbs, but others use their snouts to dig for food, to burrow, or to bore holes.

Pigs use their snouts to search for and dig up food.

Weevils feed by boring holes into all sorts of things, from rice to wood. Female weevils also bore holes and lay their eggs in the holes. Earthworms burrow by pushing their pointed front end through the earth, swallowing the soil as they go.

Bandicoots dig their pointed snouts into the ground in search of insects and worms. They wipe the worms clean with their forepaws before eating them.

The walrus uses its long tusks for digging up shellfish, as well as for fighting.

Special Uses

Stag beetles have jaws that look like antlers, and they use them for fighting in the same way as a male deer uses its antlers.

Squirrels and chipmunks collect and store food in their cheeks until they get it home, as do baboons.

The mallee fowl buries its eggs in a mound of sandy soil, and uses its beak as a thermometer to test the temperature of the mound. The mallee fowl wants to keep the right amount of heat for the eggs to develop, so if the eggs are getting too hot, it scratches away some of the soil to cool them down; if they are too cold, it piles more earth over the eggs to warm them up.

The elephant's trunk is probably the most special nose of all. With its trunk an elephant can lift a ton of timber, pick up a peanut, trumpet a warning, spray water or take a bath, pluck a flower, or caress its child.

Index

alligator turtle 15
anteater 4, 6
archerfish 11
baboon 18
bandicoot 17
bear, sloth 11
beetle, stag 18
birds 5
blue whale 12
butterfly 5
cats 5, 10
chipmunk 18
cormorant 15
dogs 10
earthworm 16
elephant 4, 20
flamingo 13
fowl, mallee 19
frog 10
gannet 14
heron 15
kingfisher 15
kiwi 8
krill 12
leopard 10
mallee fowl 19
mole 9
   star-nosed 9
pelican 14
penguin 15
pig 16
skimmer 14
sloth bear 11
snail 5
squirrel 18
stag beetle 18
star-nosed mole 9
tapir 7
turtle, alligator 15
walrus 17
weevil 16
whale, blue 12
wolf 10